Looking Back

KIRKLAND LAKE

ONTARIO

ROZA BROWN IN MOURNING. Rosie's main interest was the Royal family. The tough old pioneer friend of Harry Oakes and real estate adventurer often carried around pictures of royalty along with her ever-present pack of dogs. Here, she poses in a formal studio portrait with a picture of King George V. She is dressed formally but note the rubber boots peeking out below her long dress. (Museum of Northern History)

For Irene & Dave —
good chums — I know
this will send Irene down

Looking Back

KIRKLAND LAKE

ONTARIO

memory lane

Best from Joan [illegible]

Michael Barnes
June 14, 2003

Michael Barnes

Looking Back Press

Vanwell Publishing acknowledges the financial support of the Government of Canada through the Book Publishing Industry Development Program for our publishing activities.

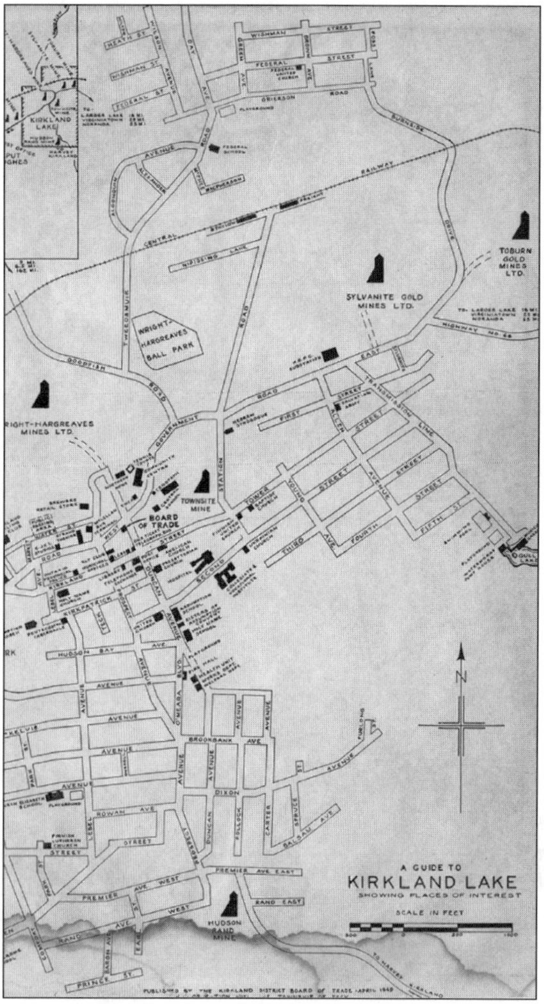

Published by Looking Back Press
An Imprint of Vanwell Publishing Limited
1 Northrup Crescent, P.O. Box 2131
St. Catharines, ON L2R 7S2
For all general information contact Looking Back Press at:
Telephone 905-937-3100 ext. 835
Fax 905-937-1760
E-Mail vanessa.kooter@vanwell.com

For customer service and orders:
Toll-free 1-800-661-6136

Printed in Canada

National Library of Canada Cataloguing in Publication

Barnes, Michael, 1934-
Kirkland Lake, Ontario / Michael Barnes.

(Looking back)
ISBN 1-55068-924-X

1. Gold mines and mining—Ontario—Kirkland Lake—History.
2. Gold mines and mining—Ontario—Kirkland Lake—History—Pictorial works.
3. Kirkland Lake (Ont.)—History. I. Title. II. Series: Looking back (St. Catharines, Ont.)

TN414.C22O5 2003 338.2'741'0971314 C2003-902715-5

Contents

Notes from Author

Today, Kirkland Lake is a quiet Northern Ontario town, but it was once a hive of activity. From the early twenties to the late sixties, the mines operated twenty-four hours a day, seven days a week. Their whistles, which signalled shift and other changes, were so reliable that residents used them to set their clocks. From the east to the west, the Toburn, Sylvanite, Wright Hargreaves, Lake Shore, Teck Hughes, Kirkland Lake Gold, and Macassa mines supported a busy, prosperous town. Sadly, only the Macassa mine still produces gold within the town limits.

Years of mining produced millions of tons of waste rock, which was pulverized in the milling process and reduced to a sand-like consistency. Called slimes or mine tailings, this unwanted by-product still covers the land in controlled, dammed-up deposits, a perpetual reminder of the time when such material signified the wealth that gold production delivered.

Children who grew to adulthood during the mining period mostly left the North to pursue opportunities in education and employment. Many of these former residents return to Kirkland Lake and marvel at the place that gold made and reminisce with their friends. Reunions are frequent and always well attended in Kirkland Lake.

Not many communities boast the fine country home of a pioneering mining multi-millionaire. But such is the case in Kirkland Lake with the Sir Harry Oakes Chateau. After its original owner left town, it was a meeting place for Lake Shore Mine company directors, and later operated for a few years as a nursing home. Today, the Chateau houses the Museum of Northern History. This is a great place to view the artifacts of past years and accomplishments of Kirkland Lakers. Visitors marvel at the wealth of mining memorabilia showcased in this historical and well-maintained building.

Curator Robin Ormerod and his assistant Linda Sinclair were most generous in giving their time and assistance in making the archives available for the selection of images used in this book. As ever, my wife Joan, nee Wyatt, who is herself a former resident of Kirkland Lake, put up with all the travel and enthusiastic history talk. The resulting work is the responsibility of the author alone.

Introduction

After the great gold strikes in the Yukon on the Klondike trail of the '98, the attention in Canadian mining circles shifted to Ontario. The provincial government commenced building a railway north from North Bay in 1902 to provide an entry into Northern Ontario for settlers hoping to take up cheap crown land for farming. Silver was found along the tracks at what would become Cobalt. The new line reached Cochrane in 1908 and by 1911 the Temiskaming and Northern Ontario Railway required a spur line to go to Porcupine and Timmins to access huge gold finds there.

Several prospectors felt that gold might be found somewhere between Cobalt and Timmins. The first gold-seekers tried the ground around what are now Larder Lake and Virginiatown in 1906 without success. Discoveries in that area would come later. Gold turned up in several places along the railway tracks. Minor deposits were finally noted at Bell's Corners, which was later called Swastika—a name whose origin is thought to be Aboriginal and not connected to any sinister German regime. At that time, the prospectors had no time to reflect on names. They began to fan out three miles to the east to a place by a scenic waterway. A motley group of people made their way to the land around and near this pretty lake, which was later given the name Kirkland.

As ever, some prospectors came to Kirkland Lake, failed to strike gold, and left with their pockets emptier than when they first arrived. Other successful claim stakers banded together for mutual support and sold out their promising properties to mining interests who had the resources necessary to develop the ground. Two men stand out from the group. Harry Oakes was an American from Maine who had roamed the world in search of a lucky strike. After many trials and tribulations, he established the great Lake Shore Mine, became fabulously wealthy, and, later in 1943, was murdered in the Bahamas in a bizarre crime that has never been solved. By contrast, Bill Wright discovered the property that became the second largest in the camp, the Wright Hargreaves Mine. The former cavalryman in the Boer War went on to serve his country in the Great War, owned The Globe and Mail, and passed away in the early fifties without an enemy in the world.

Against this backdrop came the people who worked in the mines, married and had families, and laid down roots in the camp. Although many mining areas were truly camps in the sense that mines were played out and then prospectors moved on to new prospects, Kirkland Lake developed into a thriving community and enjoyed more than ninety years of progress. For many years Kirkland Lake had a rough pioneer image. Rock was everywhere and the town was built on it. This made building expensive at times, for water and sewer services had to be blasted out of the living rock. The town became incorporated in 1919 and gradually improved community services. At first, money for mine development was difficult to obtain, but when the worth, quality, and quantity of the precious yellow metal was proved, there was no dearth of investors. As a result, seven great mines were sited within the town, close by the main thoroughfare through the community.

While the Great Depression caused much hardship throughout Canada, the gold mines stayed open at full capacity, and so towns like Timmins and Kirkland Lake prospered. By 1938 Kirkland Lake reached its peak of about 26,000 people. But then the Second Word War drew the nation's attention to the war effort and gold mining was considered a non-essential industry. Many miners and their families left town during a strike in 1941. Other workers in the industry went to places like Sudbury that required a work force to produce metals vital to the pursuit of the war effort. Kirkland Lake recovered somewhat after the war but never regained its former size or gold production. The mines inevitably declined and most closed in the first half of the sixties. This spelled apparent doom for a single-industry town, but then a huge iron ore deposit was located south of Kirkland Lake. Once again, the area began to boom. For more than twenty years the giant Adams Mine took up the economic slack left by the closed gold mines and provided a handsome living for its employees and prosperity for the town where they lived. This operation closed in the late eighties and with only one gold mine left producing ore in the community, Kirkland Lake was forced to reinvent itself again. This time it grew as a regional, educational, medical, and service centre. Tourism was high on the local prosperity agenda. Today, the beauty of the rugged northern area attracts tourists, cottagers, and others travelling through Northeastern Ontario.

No one can forget the importance of mining to the history of Kirkland Lake. Signs of its impact are found everywhere and the Museum of Northern History, the Sir Harry Oakes Chateau, is a good starting place to rediscover the days when gold made Kirkland Lake.

One
Discovery and Staking

Two billion years ago, volcanoes erupted and changed the landscape of the Kirkland Lake area. In their path to the surface, streams of boiling water deposited gold in fissures called veins. The ground was further altered and worn down by great glacial ice sheets, which ebbed and flowed across the land. So the golden treasure waited undetected for thousands of years until early in the 20th century, when prospectors began searching for the precious golden metal. A few were successful.

MAKING THEIR MARK. Government and railway surveyors were some of the first to arrive in the area not long after the turn of the century. The Temiskaming and Northern Ontario Railway line had to be chosen and the right-of-way mapped. Only one man may be identified in this 1907 view. Frank Duncan stands to the right. He stayed in the Kirkland Lake area and was a successful businessman and entrepreneur. Duncan Avenue is named for him. (Museum of Northern History)

READY TO GO TO WORK. This group of surveyors is pictured in the area of the hamlet of Swastika, the rail station most commonly used to access Kirkland Lake. They were likely working for the railway, the forerunner of the present Ontario Northland. Notice their heavy boots and the worn paddle , indicative of hard travelling. (Museum of Northern History)

THE WAY NORTH. By 1910 the tracks had reached as far north as Cochrane. One of the earliest of the smaller mines in the area was the Lucky Cross, which was an allusion to the name Swastika. This mine was close to the Swastika station and the water tower for the steam engines may be seen on the left. The Lucky Cross was not a moneymaker and today all that remains are a capped mine shaft and some acres of broken rock. (Museum of Northern History)

WAITING TO GO. These men are early drillers at one the Kirkland Lake mines. They have set up a steel post to hold up the "roof" and are drilling at an angle into the rock. Note the complete absence of hard hats or helmets. Protective headgear was not mandated until much later. (Museum of Northern History)

EARLY CAMP. Another property in the south part of Kirkland Lake was the Hudson Bay Mine. In common with so many ventures in the north, this property never made money, though it was operated by more than one company and was quite a big operation with its own mill. Some people said that veteran prospector Sandy McIntyre helped to "salt" the mine, or make it appear that there was more gold in the workings than actually existed. (Museum of Northern History)

WAITING FOR CIVIC IMPROVEMENTS. Swastika, although now a small community adjacent to Kirkland Lake, was in the early days the larger of the two settlements. People arrived by train. For many, it was their introduction to the northeast. This photograph shows the Doig store and in the centre background the railway freight sheds. (Museum of Northern History)

PROUD DISCOVERER. Ex Boer War veteran Bill Wright stands by his tent on claims which became the great Wright Hargreaves Mine.
(Museum of Northern History)

LONELY WALK. The first mine to get underway was the Tough Oakes, later the Toburn Mine. Harry Oakes and the Dusty brothers staked the property by Gull Lake, on the present site of the Kirkland Lake campus of Northern College. A lady named Russell is seen here walking from the mine to Kirkland Lake. The hydro poles indicate that electricity was received from the power generating station at Charlton. (Museum of Northern History)

WALKING THE TRACKS. Margaret MacIntosh and Agnes MacIntosh are seen here examining one of the first shafts at the Tough Oakes Mine, which had five shafts in all. Money made from the Tough Oakes helped bankroll Harry Oakes on his way to developing the Lake Shore Mine. In the early twenties, the mine was reorganized and became the Toburn Mine. The graceful headframe of that property still survives and delights photographers, painters, and visitors alike.
(Museum of Northern History)

TAKING A BREAK. Prospectors Hiram Tobico and Len O'Brien are seen here trenching and endeavoring to trace a possible gold bearing vein. (Museum of Northern History)

Two
Getting Started

Major city newspapers often refused mining advertisements, especially those offering share placements because there had been many crooked deals with regard to mining prospects in the past. During the First World War, development of the Kirkland Lake mines slowed as miners left their jobs and went to fight for their country. Among them was Wright Hargreaves Mine owner Bill Wright, who left his fledgling property on a care and maintenance basis while he went off to fight. Harry Oakes, however, stayed with his prospect. He sold his interest in the Tough Oakes Mine to concentrate on the Lake Shore Mine and worked doggedly with very little help. Once he struck spectacular gold showings under the lake itself, he knew he had found the mine of his dreams.

After the war, the industry really picked up and the rough-and-ready town of Kirkland Lake became incorporated and settled down to a time of progress. Deposits of gold were discovered and drew an influx of workers to the area, which became known as the Mile of Gold.

THE ONLY MINE IN TOWN. The Tough Oakes was the first producing mine in Kirkland Lake. To the left, out of sight of this view, is Gull Lake. There are mine buildings and workers' residences dotted among the ever-present rock. Nothing shown in this photograph exists today. The only reminder of the mine is the elegant Toburn head frame and the provincial government plaque about the Mile of Gold. It stands by a gold vault that once stood inside an office building that burned down in 1924. (Museum of Northern History)

TRANSPORT HUB. Although a small community, Swastika was a busy place because it was the destination for both freight and passengers headed for the new gold camp of Kirkland Lake. Here, a freight team is seen heading for town. The railway station, not in its present location, is seen in the centre. To the left sits Joe Boisvert's hotel. Joe catered to all comers, especially those who had limited business in Kirkland Lake and were just passing through. (Museum of Northern History)

GOLD SEEKERS. The unknown men in this undated picture are a prospecting team and they pose by a new claim post. Each claim was marked at its corners by posts squared with the axe and marked in pencil. The mining recorder issued metal tags and these were nailed to the posts. Despite the care taken in pacing out and naming claims, there were often disputes over ownership. (Museum of Northern History)

PEACEFUL VILLAGE. This is how Swastika appeared in 1912 in a view likely taken from the station. There is a church on the hillside and to the right a small but very important hardware store. Buildings were of frame construction as brick was in short supply and used mainly for chimneys. (Museum of Northern History)

HOLDING UP THE ROOF. These men at the Tough Oakes Mine are setting up for work at the rock face. One man, right, is adjusting a jack near the air and water hoses. The other is sorting out drill steel. (Museum of Northern History).

A VIEW OF WORKING PLACES. In 1915 the Beaver Mine of Cobalt (bottom left) was developing Kirkland Lake Gold Mine, the only mine bearing the town's name. This is the reason why the hill coming into town from the west is still called Beaver Hill. The great Lake Shore Mine (top) shown when it had reached to a depth of three hundred feet. La Belle Kirkland Mine (bottom right) had a sixty-degree incline shaft. This property was a dud and never made money. (Museum of Northern History)

COMING HOME TIRED. This 1917 photograph recalls a great snowstorm. This team kept the road open between Kirkland Lake and Swastika. Note the drag of timbers at the rear that held down the snow and pushed it along. (Museum of Northern History)

READY TO MUSH. This 1919 crowd outside the café and pool hall watch a form of transportation that was fast disappearing from the community. The single-harness sleigh was obviously used for fun and maybe bringing home the groceries. The two buildings to the right are on the approximate present day site of the Royal Bank. (Museum of Northern History)

THE LAKE SHORE CREW. This was the first bunkhouse for mine employees; many of whom were either single or were waiting for families to join them. Note the two cooks in this 1922 photograph. Harry Oakes (far left) sits on the verandah rail secure in the knowledge that his great mine was now showing a handsome profit. (Museum of Northern History)

A LIKELY PROSPECT? La Belle Kirkland was located at the southern tip of Goodfish Lake, which is near the present-day location of the airport. Trace gold was discovered there, and the prospect seemed good since the location was just to the north of the Mile of Gold. Unfortunately, the mine never produced profits for investors and no trace of it may be found today. (Museum of Northern History)

A MODEST START. This tiny shack held a jail and the municipal offices when Kirkland Lake was incorporated in 1919. A major civic improvement requested by citizens is shown in the form of a nice wooden sidewalk. (Museum of Northern History)

EVIDENCE OF PROGRESS. This photograph looks north on Duncan Avenue. A team is seen leaving the Wright Hargreaves Mine property. The head frame is out of view, but the water tower is shown. Streetlights were a welcome addition. The present OPP building would be somewhere between the two buildings to the left centre. (Museum of Northern History)

TOP OF THE POLE. The tame bear looking out over Kirkland Lake in 1917 could still see a rough-hewn developing community where stumps and just-cleared bush predominated. There is little evidence of town planning in the picture, which is only natural since incorporation was two years away. (Museum of Northern History)

A FLEETING VISIT. In 1919 the Prince of Wales visited Canada and toured the North by train. He passed through Swastika and waved to folks from the rear of the train while it took on water and other items. There was no time to set up opportunities for photographers and the man on the rail car (left) may be the Prince. During his Cobalt stop, he toured a mine, was plied with hospitality, and given a silver bar. The Doig Hotel is shown in the background (left). (Museum of Northern History)

GROWING WITH SERVICES. The horses pulling the big plow on Government Road, the new town's main street, would have seen continual change. Buildings were being erected, finished, and opened for business even through the winter. (Museum of Northern History)

THE THOUGHTFUL VETERAN. This portrait of mine owner Bill Wright was taken *c.*1918 as he finished his war service. He was a quiet, gentle man and popular with all that knew him. He would come back to the gold town and spend the next couple of years building the infrastructure of the second largest mine in the camp. When asked about his smoking he replied that the habit had been a lifesaver for him in the Boer war. A survivor of the siege of Ladysmith, he was one of many who were often hungry and he said the tobacco eased hunger pains. (Museum of Northern History).

SHARES TO RICHES. Harry Oakes sold his ownership in the Tough Oakes Mine to finance the start-up of his new property, which consisted of four claims adjacent to the waters of Kirkland Lake. As part of the deal he likely received this share certificate issued by the first producing gold mine in the camp. He obtained it on June 15, 1915, just four years after he arrived in Swastika practically penniless. The back of the paper shows that he sold the investment instrument on March 16, 1917. If the investment was at face value, at five dollars a share it represented $100,000 in

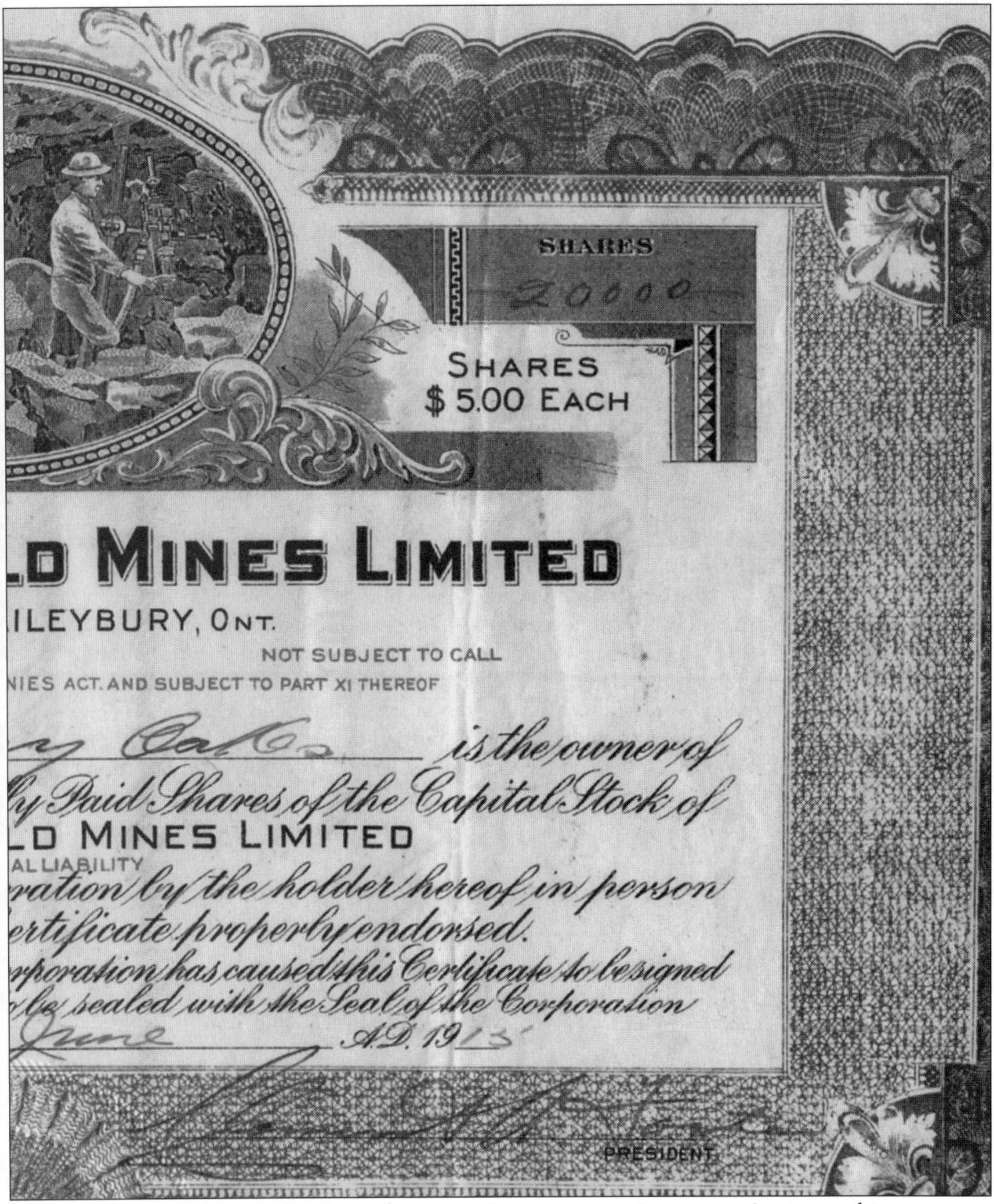

sorely-needed funds for the cash-strapped mine owner. In present day terms this would represent about six million dollars. The year Harry Oakes sold this certificate, his men struck a rich gold vein under Kirkland Lake and his financial worries were over. Mine shares were very ornate and presented to offer an impression of stability and confidence in the product. Note that the head office of the company was to the south in Haileybury. The District town of Temiskaming, it was a centre of banking and mining activity. (Museum of Northern History)

A NEAT PROPERTY. By 1920 the Tough Oakes Mine was a far cry from its early rough-hewn days. In the background is Gull Lake where Harry Oakes danced in front of a blazing fire after he had successfully staked the mine and beaten other prospectors in the race to claim the land. The road running diagonally through the picture is roughly in the same direction as the present highway that leads to Larder Lake, Virginiatown, and on to Rouyn and Noranda in Northwestern Quebec. The large building contained a gold vault. In 1924 it burned down and the solid concrete building it housed is now on display next to the historical plaque sign. (Museum of Northern History)

Three
Progress and Development

The twenties saw the consolidation and rich development of the seven mines on the mile of gold and progress in civic improvements in the community. The Teck Hughes, third largest of the properties, had a rough time when an official forgot to pay provincial taxes. When the mine manager saw two local men staking the mine's claims, the delinquent taxes were paid and the hopeful stakers were bought off. The Kirkland Lake Gold Mine was slow in finding worthwhile gold deposits until 1926 when it brought in new management. J. B. Tyrell was a famed Canadian geologist and explorer. He felt certain that the deposits went deep and once the shaft was driven to the 743-metre level, the effort paid off in high gold values. The Sylvanite Mine came into its own in 1927 and the last mine in the town, the Macassa, was incorporated although it did not achieve positive production until the early thirties.

A SCENIC SPOT. No one has positively fixed the location of this slender bridge over one of the bays of Kirkland Lake. The pretty waterway was a contrasting backdrop to the Hargreaves, Lake Shore, and Teck Hughes mines, which ringed its east and south shores. Eventually their activity would destroy the lake. (Museum of Northern History)

WAGON OF RICHES. In the early days of the Kirkland Lake mining boom, it was not uncommon to see gold bars waiting on the station platform at Swastika to be picked up by the next train south. Here, station agent Jack Gale and prospectors Bill Brennan and Tom Furlong stand hefting three of five gold bars in this 1922 memory. (Museum of Northern History)

COSY SPOT. Scores of mines were developed in the Kirkland lake area. Only a few became great producers. This is a view taken in 1922 of the manager's house at the Hunton Mine. The property, located on the edge of town, never made money. It may be found a couple of hundred yards due south of the Don Lou Motel, once Harry Oakes' bunkhouse. Only a hole in the ground remains where the house stood. (Museum of Northern History)

RELAXED TRAVEL. This snowy view up Duncan Avenue to the Wright Hargreaves Mine illustrates the usefulness of a dogsled and team in the winter months, whether for fun and relaxation or for bringing home the groceries. (Museum of Northern History)

KING OF THE HILL. In the early twenties the mine at the top of the hill on the way into Kirkland Lake was neat in appearance but not doing well in finding gold ore. Investors were always attracted by mines that included the name of the camp in their title. (Museum of Northern History)

ROUGH VIEW, NICE WALK. In 1923 this view east from Prospect Avenue along Government Road was rather bleak to newcomers. But town pioneers would point to the sturdy sidewalks as a sign of progress and, hidden behind the pole centre, the busy Central Public School. The Imperial Bank was the first financial institution in town. After a while it moved across the road to the corner of Prospect and Government and is today the Canadian Imperial Bank of Commerce. (Museum of Northern History)

THE FIRST UNITED CHURCH. One indication of the rapid growth of Kirkland Lake was the building of the first United Church in 1920, just one year after incorporation. It was located on the site of the present Arcade building on the south side of Government Road. Perhaps after the service some worshippers went down the steps to have a meal at the Coney Island Lunch. The church served until the early thirties, when the second and present United Church was built on Second Street.

A PLEASANT PADDLE. During the late twenties, folks could still paddle their way around Kirkland Lake and some men even used this convenient form of transportation to make their way to work. By this time, however, the lake was rapidly becoming polluted with mine tailings and the effluent mix of chemicals that was associated with them. In this view, the United Church is visible just up and behind the canoe. (Museum of Northern History)

A WINTER SCENE. In this 1924 view, looking east along Government Road from the approximate site of the present Esquire Cigar Store, the town looked peaceful and clean. The snow masked the debris from clearing land and erecting frame buildings. (Museum of Northern History)

REMEMBRANCE DAY. Snow must have been late coming on this particular November for the wreaths are still visible on the war memorial. The cenotaph in Kirkland Lake has occupied several locations over the years before coming to rest on the Hargreaves property opposite the Royal Canadian Legion. Here, it is shown on the grounds of Central School on a rock base that has long since been blasted away. There is no recollection of the fate of the artillery piece. Perhaps it was melted down for use in the Second World War. (Museum of Northern History)

HUMBLE BEGINNINGS. This is the first shaft house of the Wright Hargreaves Mine in the early twenties. The Lake Shore Mine may be seen across the lake. The site of the shaft house is now occupied by the liquor store. There is a thick steel reinforced cap over the former shaft, which had a depth of nearly eight thousand feet, but locals still joke that one day maybe all the bottles will fall down the shaft. (Museum of Northern History)

GOOD SIZE LOADS. Little Transportation always had enough work to keep busy. Their trucks hauled freight into the community and often brought out locally-cut logs. (Museum of Northern History)

GOING DOWN. The Kirkland Lake Gold Mine was located at the top of Beaver Hill, so named for its original hoist and mill machinery, which came from the Beaver Mine in Cobalt. This photograph was taken in 1926 when the shaft was driven deep to access low-lying gold. Nothing remains today of the buildings seen here. (Museum of Northern History)

CALM REFLECTION. In this 1926 view the Wright Hargreaves is seen across the lake from its larger rival, the Lake Shore Mine. By that time, the mine Bill Wright had founded was making him a millionaire. He had bought a house in Haileybury on Millionaire's Row, but it was destroyed in the great Temiskaming fire of 1926 and he bought new property in Barrie. Of all the buildings shown in this photograph, only one survives—the house almost centre in this picture. (Museum of Northern History)

WAITING FOR THE TRAIN. By 1926 the automobile had largely replaced the faithful oat burners. Here taxis and private cars are parked at the Swastika station. The station building and freight sheds seen here are gone now and the rock long since blasted away. (Museum of Northern History)

BROKEN ROCK. Two mines are shown in this view. A Sylvanite headframe is shown in the middle background while the bulk of the picture is of the Wright Hargreaves. This illustrates clearly the waste rock that was produced from the mine and the way the property continually encroached upon the waters. By 1929 the once scenic lake was completely gone, filled in by rock and mine tailings. (Museum of Northern History)

GREAT ROOMS. This view looks north across Government Road to the Lake Shore Mine. The long building in the foreground was the bunkhouse Harry Oakes built for his single miners. It had a dining room and good accommodations. Later Harry had a large greenhouse added to the end of the building to supply fresh vegetables for the kitchen. The present mall covers the large mine mill buildings seen in the rear centre of this photograph. (Museum of Northern History)

GROUP SHOT. The miners of one shift who went to work at Lake Shore Mine using Number One shaft are shown in this 1926 photograph. Rail lines, shown in the right foreground, were used to move ore cars in and out of the shaft house. On the left, one joker has placed his lamp on the head of the lad in front. The names are lost to us. (Museum of Northern History)

A NEAT OPERATION. The Sylvanite Mine did not really get underway until 1927. The first headframe is shown here. A hydro station now covers much of the ground seen in this picture. The location is opposite the present-day Masonic Hall.

BUSY STREET. Government Road was under constant change in this 1927 view. Lumber lay everywhere as buildings were continually improved and built. It was said during this period one could never move far from the sound of carpenters' hammers. The Strand Theatre may be seen in this photograph (left centre). The water tower that gave its name to Tower Street is on the skyline. (Museum of Northern History)

WORKING MINE. In this photograph, the Sylvanite Mine headframe has been covered with white tarpaper. The hoist room is on the right. Ore hoisted through the shaft would go onto the conveyor to the mill at the left. (Museum of Northern History)

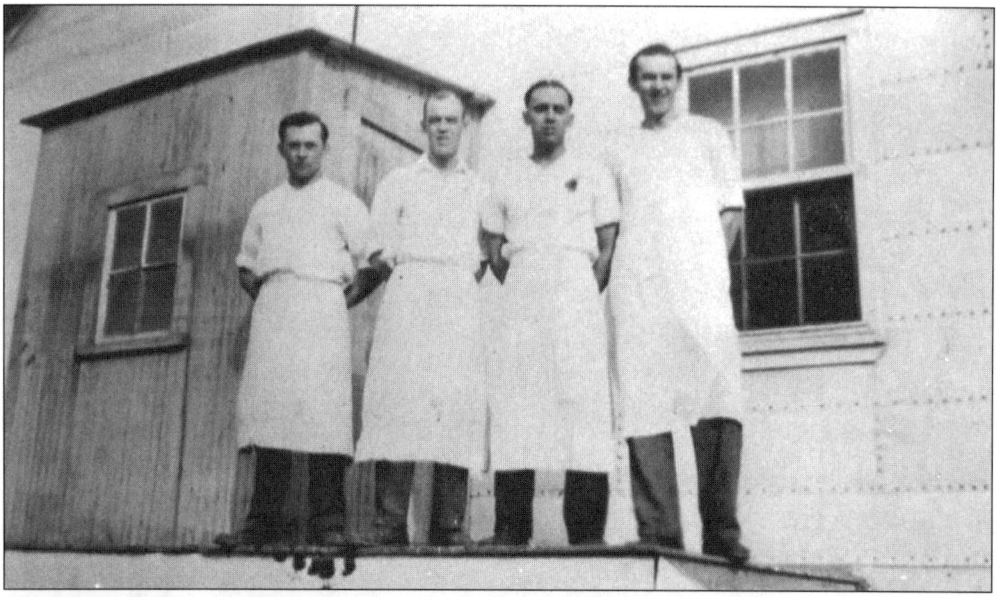

WHITER THAN WHITE. These cooks at the Wright Hargreaves Mine served meals around the clock. Their names have not survived. The mine discontinued the practice of serving meals once there were accommodations available for workers in the town. (Museum of Northern History)

A MODEL MINE. In 1927 the Wright Hargreaves Mine workers could see their Lake Shore Mine counterparts across the lake, and to the far right the Teck Hughes Mine. Ore travelled up the incline conveyor to the mill. Today, tennis courts are located in the area to the left in the foreground. The author's father-in-law, George Wyatt, ran the hoist in the building shown at right centre. (Museum of Northern History)

ROUGH ROAD. In this view, the business section of Kirkland Lake appears to have a wide thoroughfare, but in fact there were rocky outcrops along each side of the street. Over time these were blasted away. Charlie's Hotel is shown near the car on the left. (Museum of Northern History)

LABOUR INTENSIVE JOB. The local newspaper started in Cobalt and then moved north to Kirkland Lake. In the early years of its operation, the paper only came out three times a week. Typesetters are at work setting the print pages with presses (right). The newspaper office still occupies the same building. (Museum of Northern History)

GOING DOWN. This Sylvanite Mine shaft had two compartments. The left was for ore, waste rock, and materials. The smaller compartment was for men only. This was a relatively small shaft. Note the poles over the opening to prevent injury from falling rock and the drill steel (right). (Museum of Northern History)

EDUCATION GAINS. In 1928 the newly built school flanked the old Central School. By the late twenties school enrolment was soaring and classes were large. Newcomers built homes and settled in with their families. (Museum of Northern History)

RAIL LINK. By 1928 Kirkland Lake had its own station as the Temiskaming and Northern Ontario Railway built a line from Swastika into Quebec and connected with Rouyn and Noranda. The electric car was a great clean innovation but sometimes the batteries failed and a steam train would come and give it a tow. (Museum of Northern History)

BIG COMPLEX. The third largest mine in the camp, the Teck Hughes, stood at the bottom of Beaver Hill. Senior staff lived in large houses, shown in the foreground. Eventually there were headframes on either side of the road and a conveyor ran across the main highway with a Welcome to Kirkland Lake sign on the siding. (Museum of Northern History)

NICE FACILITY. Frederick Banting, co-discoverer of insulin, opened the Red Cross Hospital in 1926. It was supported by the mines and had an adjacent residence for nurses. (Museum of Northern History)

HARRY'S DREAM. In 1929 Harry Oakes' Lake Shore Mine was the largest property of the seven Kirkland Lake Mines. The mine employed two thousand people. The mine owner soon ceased to live in Kirkland Lake and occupied himself with his other homes in New York, Bar Harbour, Maine, and London, England. (Museum of Northern History)

TRAIN ON TIME? Although there was a rail line through Kirkland Lake to Quebec, the majority of the passenger traffic still came north through Swastika. Meeting the train was always a popular pastime. This is the first station building. The present station is to the south of it. (Museum of Northern History)

OFFICIAL VIEW. In 1929 the Geological Survey of Canada went round the north recording pictures of mines for posterity. The resulting record shows the extent of the Wright Hargreaves Mine but does not indicate that Kirkland Lake had become unrecognizable as a lake, having been filled with tailings. (Museum of Northern History)

PARDON THE ROCK. In 1929 the Strand Theatre was so popular that it featured shows all day to meet the needs of miners on shift work. The unsightly rock in front was blasted away and eventually the theatre was enlarged. The road was not paved and could become quite muddy in spring. (Museum of Northern History)

A MURKY ASPECT. This 1929 photograph is taken across the slime or tailing filled lake to the southeast. When dry, the tailings gave the appearance of a sand desert. A car could travel over the surface. . This photograph shows the Lake Shore Mine and behind it the Teck Hughes Mine. (Museum of Northern History)

UNCOVERED FRAME. Mine headframes were huge buildings that reached high into the air. Skip and cage cables were suspended from them down into the deep mine shaft. Here, the Sylvanite shaft house or headframe was being sided. Note the man at the top of the structure. When mines closed, the steel skeletons were dismantled and erected at other properties. (Museum of Northern History)

FENCED IN. Security was important for the mines. Fences surrounded the property and equipment. Security staff guarded gold ore and bars. (Museum of Northern History)

Four

Depression and Boom

The Great Depression hit Canada in 1929 and held the country in its grip for about five years before the economy gradually began to pull out of the worldwide slump. In common with other municipalities, Kirkland Lake struggled to provide welfare for those out of work. Soup kitchens provided food and the town offered employment a few days a week for those without regular work. Eventually the town defaulted on its obligations, as did so many other communities, and it was supervised by provincial government regulators. In spite of these difficulties, the mines went on working around the clock. Unemployed men from across the country came seeking jobs and their influx added to the problems of mining towns. There were always men waiting for jobs at the mine gates. Eventually Kirkland Lake recovered in the late thirties and the population boomed, reaching 26,000 people in 1938.

CLOSE WATCH. A worker pours molten gold in the closely guarded refining room. For many years workers did not wear protective gear such as asbestos gloves, face shields, and heavy-duty coveralls. The gold in this mould was cool enough to pick up within half an hour of the pouring process. (Museum of Northern History)

TITLED PROSPECTOR. Freeman Mitford, Lord Redesdale is seen here on his claims, not far from the junction of Highways 66 and 112. He spent several summers working on his claims and was by no means a remittance man or a poor, unwanted member of a rich family, as many locals thought. He is better known for his famous daughters. Diana married Sir Oswald Mosely, Britain's top fascist, Unity was a follower of Hitler, and both Jessica and Nancy became well-known writers in the United States. (Museum of Northern History)

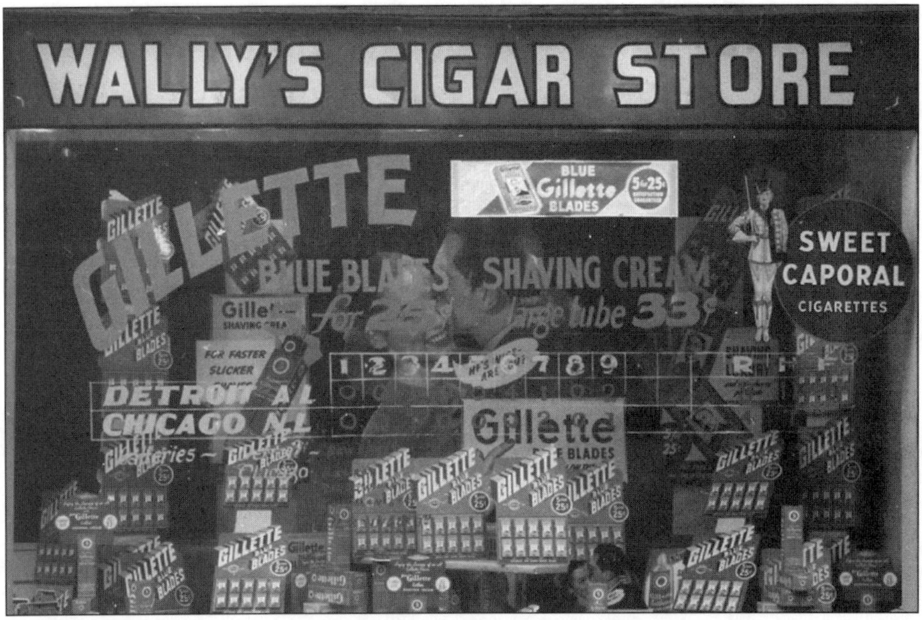

SMOKE SHOP. Although this window advertises cigarettes, the display was almost exclusively devoted to a razor blade promotion at this popular store. Note the scorecard for hockey games so that the results could be pencilled in as they were announced. (Museum of Northern History)

MODERN RIDE. The Lake Shore Mine adopted all modern innovations. Here, a miner drives an electric train. The battery-run machine could pull ore cars along the tracks. An advantage of this form of transport was that electric automotive power did not pollute the precious air underground. (Museum of Northern History)

MINE AND TOWN. This view looking southeast from high up on Lake Shore Mine shows houses built by Harry Oakes for his employees (rear centre row). These properties survive today. In 1931 the town did not yet extend far to the south. Parallel to Government Road, elevated tailings discharge pipes may be seen. The building to the left of the headframe was the impressive mine office. (Museum of Northern History)

CLEAN FLOOR. The Lake Shore Mine compressor building was neat and well kept. Mines used vast amounts of both electric and compressed air power and it was important that these machines operated to peak efficiency. (Museum of Northern History)

BY LAKE EDGE. The Teck Hughes Mine was practically a town site in miniature. Many employees lived on the property, which was only a short walk from Kirkland Lake. Today, almost all these buildings are gone, but two have been preserved and rebuilt as a funeral home. (Museum of Northern History)

LONG WAIT. During the Depression, there was always a line of men waiting at the Wright Hargreaves Mine in hopes of landing a job. Here, the head mine security officer takes note of their order in line. (Museum of Northern History)

SOLID BUILT. All big mines had assay laboratories where the grade of minerals, mostly gold and silver, could be checked. This aided in milling the ore and preserving quality control. The assay building of the Teck Hughes Mine was an impressive structure and still survives. (Museum of Northern History)

FROM THE ROCKS. A rocky outcrop south of Government Road is known locally as the French Rocks. In this 1931 view the Lake Shore Mine appears in the foreground, Teck Hughes is behind it (centre), and the Kirkland Lake Gold Mine is visible in the background (Museum of Northern History)

SPOT DANCE. Not far to the east of Kirkland Lake, Crystal Beach was much appreciated by the young set. There was a popular dance hall there and families would go to relax on the beach on Dad's day off from work. Later, the Lake Shore Mine brought sand by rail from a quarry in the area to backfill its underground excavations. (Museum of Northern History)

GOING UP. Macassa Mine was the last of the seven mines to be constructed on the Mile of Gold. This 1932 view shows the first headframe nearing completion. Today it is the only surviving mine in the town. (Museum of Northern History)

READY TO GO. All mines had rescue teams ready to deal with all sorts of underground emergencies. The Hargreaves teams, shown here from left to right, in full gear in 1932, were Stan Reynolds, Alan Jeckell, John French, Chic St. Jean and A.F. Mitchell. (Museum of Northern History)

RAIL CONNECTED. This view shows the railway line from Swastika to Quebec at the top of the photograph. The spur line shown went to the Lake Shore Mine and also dropped off timber at this island in the mine tailings or slimes. From there, the wood went underground to serve various needs of the huge operation. As much as three million board feet was used annually below surface in mine operations. (Museum of Northern History)

INSPECTION. Miners worked in pairs underground with minimal supervision once they had their instructions for the day. (Museum of Northern History)

REFLECTION IN SLIMES. The Lake Shore Mine is shown here from across the tailings. Surface water pools on the sand-like lake fill. The photograph must be prior to 1938 because the great smokestack is not in evidence. (Museum of Northern History)

KIRKLAND'S FINEST. In 1932 many prominent citizens volunteered as firefighters. The Reeve and Council, also shown in the picture, were highly appreciative of these volunteers. The highly organized department even had its own mascot, lower left, and Chief Mathieson was widely respected in the Northeast.

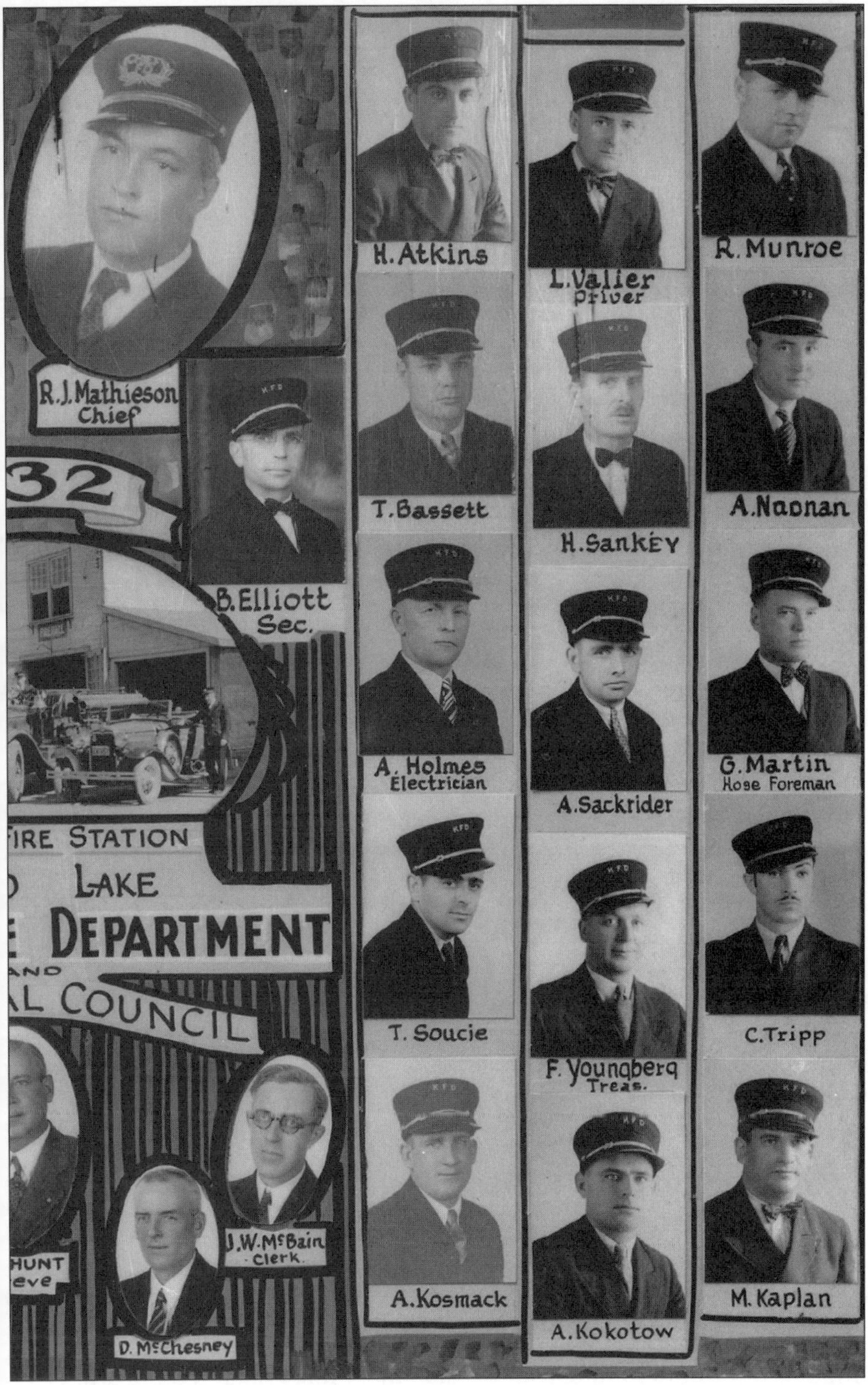

J. W. McBain was the long-serving Clerk to the municipality. Pictured next to him is Dave McChesney (centre bottom), an employee and highly-valued confidant of Harry Oakes. (Museum of Northern History)

DAY OFF. Although the Ferguson Highway, now Highway 11, connected the North from Timmins down to Toronto, the train was often the preferred means of travel for those out for an excursion. Here, a party of schoolteachers and young friends set out on a trip from Swastika. (Museum of Northern History)

TIMBER EATER. It is difficult to comprehend the amount of timber that went down into the mines on an annual basis. Used for roof props, framing, construction, and many other uses, the wood was a most necessary and significant part of the Lake Shore Mine budget. Note the spire of the French Roman Catholic Church (far left). (Museum of Northern History)

HOIST BELOW. Mine hoists had cables from the surface down to a certain depth and then separate hoists, called winzes, were installed underground to further lift men and materials from below that point. Here winze hoist man J. Germa operates at the 2,475-foot level of the Kirkland Lake Gold Mine. (Museum of Northern History)

MINE DOMINATED. The mill of Lake Shore Mine (foreground) overlooks Government Road, mine tailings, and a bunkhouse, complete with a greenhouse that provided fresh greens for the table. Tim Hortons, McDonalds, and the Veterans Affairs building now occupy what was once a tailings pond. (Museum of Northern History)

POPULAR SPOT. Gertrude Oakes, Harry's sister, supported him when he was struggling to get established as a prospector and mine owner. She was a company director, living first on the mine property and then she built a big lodge, the Red Pines, at Kenogami in the mid-thirties. The massive lodge fireplace was built out of gold-bearing rocks from the mine and the lustrous yellow metal was visible as the fire warmed the rock. Sadly, Gertrude was lost at sea when a collision sank her vacation liner off the coast of New England. (Museum of Northern History)

HILL TOP. Kirkland Lake Gold Mine is shown here at the top of Beaver Hill in the mid-thirties. The small headframe was subsequently enlarged as the mine went deeper in search of more gold feed for the mill. (Museum of Northern History)

SOLID STRUCTURE. In 1934 the mine office was an impressive brick structure. There were flowerbeds in front and the looming hulk of the mine itself behind. Today the Canadian Tire store sits on this site. (Museum of Northern History)

PROMISING LOCATION. Roza Brown, real estate owner, friend of Harry Oakes and eccentric woman-about-town, had a knack for buying up prime property. In 1934, this spot held a tailor shop, Roza's house, and behind, some said, a location for ladies of ill repute. In 1938 it became the location of the Kresge building. (Museum of Northern History)

CHANGE ROOM. During the thirties, the Lake Shore Mine employed approximately 2,500 employees. Several hundred men worked underground. They changed clothes here and showered when they came up to surface after the end of the shift. Note the Bradley washbasins down the centre aisle. Men placed their street clothes in baskets and these were hung on wires in the roof for safekeeping. (Museum of Northern History)

BIG PLANT. At one time Lake Shore Mine was the largest gold producer in Canada. Over the years it vied for that honour with the Hollinger Mine of Timmins and the Kerr Addison Mine of Virginiatown. Note the elevated plank sidewalk. (Museum of Northern History)

BIG WHEEL. At the top of the headframe, huge sheave wheels carried the great cables that hauled men and materials from deep underground. The wheels dwarf the man at left. (Museum of Northern History)

SOLID GROUP. Kirkland Lake had a strong police force in place during the mining years. Walter Little became the Federal Member of Parliament for the riding. Siegfried Atkinson was a pioneering magistrate who served various northern locations. The department was thoroughly up to date and included a motorcycle and

a system of lights on poles which, when lit, informed patrol officers that they should call the office for instructions. The police station was in the basement of the municipal building, which also held offices, Council chambers, and the courtroom. This building stood at the location of the present parking lot corner of Kirkland and Prospect streets. (Museum of Northern History)

BRIGHT SPOT. The mines worked nonstop, and whistles blew at all hours of the day and night. The lights from the gold fields illuminated the area for miles around. Eventually, people grew accustomed to the noise and lights and they could sleep soundly in spite of all the disruptions. (Museum of Northern History)

FARSCAPE. The slimes that filled Kirkland Lake run from left to right in the upper part of this picture. In the foreground lies the Teck Hughes Mine, which straddled both sides of Government Road. The hump in the conveyor belt housing is visible crossing the main road. Lake Shore Mine is in the background (upper right). (Museum of Northern History)

DUMPER. In this 1936 photograph special rail hopper cars dump Crystal Lake sand into a surface raise, or tunnel bored to ground level, for distribution underground. Despite constant back filling of excavations, there was continuous rock movement underground and collapses of workings. (Museum of Northern History)

GOOD DIGS. Harry Oakes built these homes for employees in 1936. They are large townhouses and were well constructed, surviving as apartments today. (Museum of Northern History)

GREAT HOUSE. By 1936 Harry Oakes had left Kirkland Lake, and from then on his house, the Chateau, was used mainly for mine directors' meetings and other functions. Although it was regarded as a gracious home, it must be recalled that those living in the house would have been subject to light and noise resulting from the operation of a great industrial complex right next door. (Museum of Northern History)

FINE VIEW. The Wright Hargreaves Mine stands out in this view north across Government Road from the French Rocks. The building on the left is Kirkland Lake's fourth and last movie theatre. The Uptown Theatre is now closed and is occupied by a bowling alley and apartments today. (Museum of Northern History)

ALL LIT UP. The municipal offices of the township of Teck were illuminated in honour of the coronation of King George VI. The official name of the place from 1919 until January 1st 1969 was the Township of Teck. Kirkland Lake was the town within it. By 1969 the Township was renamed Kirkland Lake and there was a Mayor instead of a Reeve. (Museum of Northern History)

JOB COMPLETE. In August 1936 the Wright Hargreaves Mine completed excavating its number five underground winze or hoist. The men pictured here sank the internal shaft to a depth of 1,420 feet and the outside measurements of this vertical shaft were twenty feet by six feet. Shaft sinking can be dangerous work and the mine recognized the quality and speed of seven months by which the work was complete. (Museum of Northern History) From left to right, as follows: (first row) A. Wright, P. Maloney (Leader), J. Myelde, J. Elliott, L. Laasko, W. Pearson, O. Proulx, A. Salo, and E. Bertils; (second row) J. Lahti, J. Laing, J. Gillies, W. Jenson,

R. Nakonechny, A. Ferguson, T. Julien (leader), G. Gauthier, O. Turgeon, and J. McDonald; (third row) W.T. Sampson (Mine Supt.), A. Soini, F. Keskinen, O. Sjostrom, G. Hawley, D. Campbell, G. Hookman, G. Dickenson, A Wiita, N. Sawchuk, E. Conroy, H. Pakkala, and F. Harrison (Shaft Captain); (fourth row) J. Antilla, W. Carlin, E. McElwain, J. Pearson, W. Julien (Leader), A. Radul, A. Proulx, N. Dale, D. McDonald, and O. Daoust.

SMART POST OFFICE. In 1936 Kirkland Lake received a fine new federal building. A clock was placed in the round spot on the roof. The basement housed armouries and the Algonquin Territorial Army unit. In recent years, when the government built a new post office on Government Road, the town bought the building and turned it into a fine municipal office. The flag is gone now; but the clock remains, although it never seems to work. (Museum of Northern History)

ACROSS TOWN. This photograph was taken looking west from the spire of Assumption Church. The parish school is visible in the far right. It is long since demolished. Kirkland Lake Collegiate and Vocational Institute is shown east of the Catholic School. Many of the large buildings on Kirkpatrick Street were large boarding houses for mine workers. (Museum of Northern History)

TOP BEAR. Illegal today, at one time bears were kept chained at service stations as an attraction for those who might stay and spend money on more than gas. Often viewers bought bottles of pop for the bruins and quite naturally this ruined their teeth. When this practice was outlawed, the animals were never again on display unless in a zoo. (Museum of Northern History)

KIRK CHURCH. The Presbyterian Church on Duncan Avenue was a welcome addition to Scots and other adherents of that faith. The church survives in the same location today. (Museum of Northern History)

NEW SHAFT. By 1938 the Macassa Mine had made so much progress after its initial slow start that it was able to sink a second shaft. This was on the aptly-named Goldthorpe Road on the site of the modern mine mill complex. (Museum of Northern History)

FULL LOT. The parking lot of the Lake Shore Mine does not indicate the number of workers busy at the huge plant. Many employees simply walked to work, as homes were not far from the property. (Museum of Northern History)

HIGH SIGN. The building that housed the tailor and jeweler remains next to the former Kresge store, but the original stores and the big sign are long gone. (Museum of Northern History)

FRAME IT. The addition to the Macassa Mill on the west side of the Goldthorpe Road in 1938 took a great deal of sawn lumber to fill in the steel skeleton. (Museum of Northern History)

READY TO GO. The miner at this unnamed property could see things much more clearly with a new rechargeable battery-operated light pack. These innovative new lamps made all the difference in illumination underground. It is hard to believe that the miners in Kirkland Lake had to make do with candles and then carbide lamps before the invention of the new light packs. (Museum of Northern History)

SMILE OF SUCCESS. Sir Harry and Lady Eunice Oakes are seen here strolling in London's Green Park in 1938 after he had received a knighthood from the King at Buckingham Palace. The once penniless wanderer had become a multi-millionaire. He donated $400,000 to St. George's Hospital at Hyde Park. (Museum of Northern History)

PRIME CORNER. This intersection was a hot business location. Eaton's was across the road from Kresge's. Horse-drawn dairy wagons plied their trade and cars competed with them for the right of way. Roza Brown had finally sold her prime corner lot and local people had some good shopping competition. (Museum of Northern History)

CHOICE ROOMS. The hotel on Government Road was a social centre popular with mining men. Its popular nickname was "The Ash Can" after owner Jimmy Ash. The man seen here washing windows is F. L. Todd, the hotel manager. (Museum of Northern History)

TOP VIEW. This aerial picture shows the Lake Shore Mine complex and Kirkland Lake. In the background is a reminder of the nearby bush that encircled the town. (Museum of Northern History)

Five

War and After

The advent of the Second World War brought troubling times for the gold mines. In 1941 a strike in Kirkland Lake closed the mines and during that six-month period, miners began to leave town. Iron and nickel mines were booming due to a need for strategic war materials. Many left for Sudbury and the waiting jobs there, never to return. The strike gained little for the workers and when it came to an end it was announced that gold would be considered non-essential to the war industry. This meant that labour would be hard to obtain, while equipment and supplies would practically dry up for the duration of the war. The mines carried on at limited capacity and their machine shops produced items for the war effort. Peace came and the mines went back to full production, but the population never did come back to post-war levels.

BIG CROWD. Kirkland Lake's pride, the Lake Shore Blue Devils, won the Allan Cup in 1940. The region went wild, as local men had taken the premier trophy in amateur hockey. This photograph was taken in New Liskeard as the victorious team travelled home. It was common for crowds to gather at towns on the way back to Kirkland Lake. (Museum of Northern History)

MULTI-IMAGE. Harold Grover, an engineer at the Toburn mine, was a fine amateur photographer. He took a picture of the former Tough Oakes Mine, forerunner of his employer, the Toburn Mine, and set four typical underground images around it. This photograph graced his Christmas card and it was also used on the cover of the *Northern Miner*. (Museum of Northern History)

FRIENDS ON THE ROCKS. Roza Brown had known Sandy McIntyre ever since he came to the camp. Sandy was a prospector. In the early days of the Timmins gold rush, he found the mine named for him but squandered the money he made. In Kirkland Lake, he had another chance at the Teck Hughes Mine, where he was hired to find the surface gold seams. Again he frittered away his money. He never had a big chance again. (Museum of Northern History)

LATE DEVELOPER. In 1940 the Macassa Mine was coming into its own and producing some of the highest-grade gold in the camp. The Macassa, located at the far end of the Mile of Gold, was the last to get started and the longest-producing mine in the community. (Museum of Northern History)

COACH. Colburne "Doc" Ames came to Kirkland Lake as a young druggist in 1928. After two years he quit his profession and went to work as a first-aid man at the Lake Shore Mine where he could make more money. An avid hockey player, he became a coach and helped the Blue Devils win the Allan Cup. (Museum of Northern History)

LITTLE ACTIVITY. This picture was likely taken during the 1941 strike as there is no smoke from the Lake Shore stack and there are few signs of activity. The Lasalle Theatre is visible in this photograph as well (centre right). (Museum of Northern History)

GOLD POUR. The miners received safety visors and protective goggles at last. The refining to process took up to twenty-six steps from the raw rock to the finished product. A trip to the refining room was always considered a treat for special visitors. The man at the top had let some gold run over the side of the mold and this would have to be retrieved for another day. (Museum of Northern History)

ORDERED CLUTTER. The Lake Shore Mine machine shop was always busy. Both surface operations and underground mining always had equipment needing repair and special items were made on order. (Museum of Northern History)

LAUNCH. In 1943 Kirkland Lake residents proudly reacted to the news that a frigate would carry the town name in the Battle of the Atlantic. It was a breezy day and as dockyard workers looked down, the dedication party held onto their hats. The ship was launched on April 27, 1944, at Quebec City. From left to right: Mrs. W. C. Rumball, unknown, Mr. & Mrs. W. Little, priest unknown, Reeve Ann Shipley, Mrs. Robert Armstrong, Mrs. William Brennan, and Mrs. Hugh MacDonald. (Museum of Northern History)

FRAMED. The picture in Roza Brown's right hand is of herself when she gained entry to the welcome for the King and Queen when they arrived in Canada in 1939. (Museum of Northern History)

READY FOR CONSTRUCTION. Men were working on the lot Roza Brown sold to Kresge's in April 1939, while across the road the Eaton's store would soon be erected. Rosie thought she had been bested in the deal and threatened to sue all involved, but it came to nothing. (Museum of Northern History)

FORMAL PORTRAIT. Harry Oakes left Canada in the thirties when taxation grew too great a burden. He settled in the Bahamas and did many good works there. He was murdered at his home in 1943 and the crime has never been solved. He is buried in Maine. Oakes was no longer involved with the Lake Shore Mine at the time of his death. (Museum of Northern History)

MOVING ROCK. This picture is taken at the 2,200 foot level of the mine, where rock is being moved to a loading area to be crushed underground before being skipped up to surface and then to the mill. (Museum of Northern History)

GONE FOR THE DAY. Miners shown here have just come to surface on the cage. They have been underground the whole shift and now head to the change room to shower and change out of their work clothes. (Museum of Northern History)

READY FOR SEA. Officers and some of the men of HMCS *Kirkland Lake* pose before setting off on convoy escort duty. None of those present are identified but one, John Black Aird, who became Lieutenant Governor of Ontario many years later. (Museum of Northern History)

PICTURE WITH AUNT. Roza Brown is pictured here with her New York dentist and nephew Dr. Bernard Saltzer in 1946. The man in the portrait is unknown. (Museum of Northern History)

BUSY PLACE. Government Road was busy as the economy picked up after the war. The town still had a frontier air about it. Many of the beer parlours had batwing swing doors, just like those in the cowboy shows. (Museum of Northern History)

LOADING MUCK. In mining parlance, newly-blasted ore at the rock face is called muck. This miner is gathering the rock in a mucker, a machine that will load the broken rock into ore cars behind it. (Museum of Northern History)

KEEP ROLLING. Large chunks of rock are gradually reduced in size in the mill so that they may be brought to the gold refining process. Here, great ball mills—revolving drums filed with iron balls—grind the rock down. (Museum of Northern History)

ON SHIFT. The miners seen here at the Toburn Mine in 1949 are not identified. The men in white shirts could have been office workers such as timekeepers or been employed in the mine dry or change room. Some miners have their electric lights

switched on. This mine, formerly the Tough Oakes and the first to operate in the camp, closed within four years. The Toburn headframe, now clad in red steel, may still be seen at the east end of town. (Museum of Northern History)

NOVEMBER 11. Roza Brown and her nephew take part in the Remembrance Day service in 1946.

STEAMER. The railway, whose name changed to Ontario Northland Railway in 1946, was a vital north-south transportation link. Here, Engine 601 hauls a passenger train in 1948 at Swastika. Over the next ten years the steam engines were phased out and replaced by diesels. (Museum of Northern History)

Six
Slowdown

As the North recovered from the war years and settled down to building up the economy once more, the fifties began as normal business times in Kirkland Lake. The mines were working and there was a general air of optimism. But underlying it all was the inescapable fact that resource industries like mining have a limited shelf life. The mines were running out of ore. Inefficient mining practices from past years meant that most mines had areas that were not backfilled, creating empty pockets where rock might shift and move, creating air blasts and causing damage underground and tragic death for some miners. The Toburn Mine closed in 1953 and by 1968 all the mines on the Mile of gold were closed except for the Macassa. Slowly, the town population began to fall.

MINE WORKS. In the early fifties the mines were all producing gold. The Lake Shore Mine parking lot is full in this photograph. To the right are houses that Harry Oakes built for senior miner personnel. Kirkland Lake had a settled look with nicely kept homes and gardens. (Museum of Northern History)

FLAG FLYING. The Anglican Church flag flew proudly next to St. Peter's on the Rock Church on Kirkland Street. In the early days of the gold camp, Harry Oakes donated a rocky lot overlooking Central School grounds to the community's first Church of England. Just before he left town forever, the mining magnate sent the deed to the property. (Museum of Northern History)

WORKING DAY. The number two shaft and headframe of the Macassa Mine brought up a high grade of gold and anchored the east end of the Mile of Gold. In the fifties, Macassa had a bright future and had no trouble hiring miners. (Museum of Northern History)

ASH CAN. The Kirkland Lake Hotel was still popularly called the Ash Can after its original owner, but had undergone improvements over the years, including incorporating the former verandah as part of the building. Two decades earlier, famed writer Gordon Sinclair used the hotel as a base as he reported on northern events. (Museum of Northern History)

SPLASH. Culver Park at Swastika was a great place to relax at the beach. Colonel Culver, a mine owner, had deeded the land to the municipality in the early mining days. It has been upgraded ever since. For a while the waters were closed due to cyanide pollution from the old Crescent Mine, but this was eventually cleaned up. (Museum of Northern History)

RESIDENTIAL AREA. The south end of town shows neat streets and many houses with nice gardens. In the seventies, the author had the honour of serving as principal of Queen Elizabeth School, the large building to the right. Pictured at the top are the former Assumption School, Holy Name School, and Kirkland Lake Collegiate and Vocational Institute (KLCVI) behind them. (Museum of Northern History)

NEAT TURN OUT. There has always been great production rivalry between the mines and also existed between their rescue teams. The Kerr Addison Mine contingent is seen here in 1955. From left to right: K.Webb (captain), M.Schaetzler, H. Armstrong, W.D. Roulston, J. Stewart, and G. Ryan (vice captain). (Museum of Northern History)

SMALL PRODUCER. Another mine outside Kirkland Lake which employed local men was the Upper Canada Mine at Dobie. Television actress Sara Botsford was the daughter of the mine manager and attended KLCVI. (Museum of Northern History)

TREE FRINGED. The Teck Hughes was the third-largest mine in town and stayed open the longest, until 1968. Here, its conveyor belt crosses Government Road. Today, the mine is gone. A Journey's End motel is located just to the right of the trees and a funeral home now occupies two former mine buildings. (Museum of Northern History)

TRAVELLING WEST. Kirkland Lake Gold Mine and mill are visible in the upper right of this picture. The Teck Hughes assay office that survives to the present is to the right. (Museum of Northern History)

SOLID PRODUCER. Bill Wright staked the Sylvanite Mine in the early days of the gold rush, but it did not get going until the late twenties. Shown here in the late fifties, it was completely demolished and little trace of the plant remains today. (Museum of Northern History)

GREAT DIVIDE. The Arctic Watershed is located some distance north of Kenogami on Highway 11. This is the place where waters run north to the Arctic and south toward the Great Lakes. (Joan Barnes)

CLOSING DOWN. In 1960 Kirkland Lake Minerals, formerly Kirkland Lake Gold Mine, ceased operations. The mine on the edge of Government Road at Beaver Hill had delivered 1,172,995 ounces of gold at a value at $35.00 an ounce with a total of nearly forty million dollars. (Museum of Northern History)

SIMILAR SCENE. Apart from the cars, this view could be much the same today as the buildings remain. The mining recorder's office to the right has closed and moved east to a new government building. The Park Lane hotel still stands but operates now as an apartment building. (Joan Barnes)

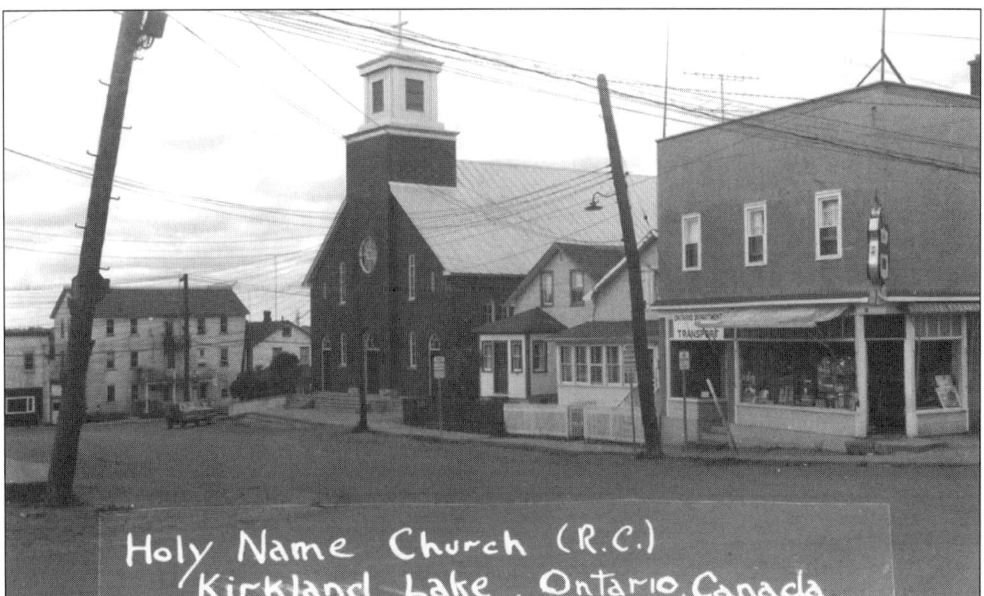

SLOPING POLES. The Holy Name Church still serves the English Catholic faithful of Kirkland Lake. The stores have changed hands but the view is similar to the past, except that the hydro poles have been renewed. (Joan Barnes)

SEEN BETTER DAYS. The Crescent Mine on the edge of Culver Park at Swastika had originally been part of the Lucky Cross Mine over the river. It never made any money. Torn down now, it was ignored by investors who moved farther east to try their luck in Kirkland Lake. (Museum of Northern History)

TALL SCHOOL. Federal School was one of the schools in the community during the mining years, but it was replaced in the early eighties by a larger building. The area shown in this photograph was called Townsite because it consisted of new subdivisions erected after the war. (Joan Barnes)

SKI KIDS. Queen Elizabeth School had an enthusiastic group of youngsters who learned how to ski at Raven Mountain. The names of students and teachers in this photograph are unknown. (Museum of Northern History)

POPULAR CORNER. The intersection of Government Road and Prospect Avenue, as shown here, is quite different from its present view. The Bavarian Tavern and Hughie Boyd's drug store are long gone. The Princess Hotel has made way for a parking lot. (Museum of Northern History)

Seven
Recovery

Gradually, the mines in Kirkland Lake closed down. Proud landmarks like Lake Shore's giant smokestack, familiar headframes, and other buildings were torn down. Much of the customary noise of machinery in various locations and the routine mine whistles marking off the hours were now all strangely silent.

However, there were good signs on the horizon. Northern College of Applied Arts and Technology in Timmins developed a large satellite campus in the old Central School. Eventually, the town built a large complex for the college on the site of the former Tough Oakes and Toburn Mines. Students were attracted to the new facility from Canada and around the world. Rumours of significant iron ore deposits south of Kirkland Lake proved to be true and the resulting Adams Mine provided close to a quarter century of employment for four hundred people. Kirkland Lake experienced many changes over the years, but local people once more looked to the future with optimism.

PROSPECTOR'S CABIN. Once there were many old log shacks like this across the Blanche River at Swastika. They were remnants of the early mining days in the area. Most are gone now, and Fireman's Park is a pretty spot for a family picnic in this location. (Museum of Northern History)

NO PITS. The Adams Mine was developed in 1964. A spur line was built in from the Ontario Northland Railway and the natural gas company added a spur pipeline to bring in a then inexpensive power source. The plant and infrastructure were built first, and the big open pits came later. (Museum of Northern History)

JOB OPPORTUNITY. The new Adams Mine was a great opportunity for newly unemployed miners and other tradespeople to gain jobs and stay in the North. This view shows the extent of the plant, crusher, mill, and loading area. (Museum of Northern History)

VOLUNTEER. The late Don Frisby was an overachiever who gave much to his hometown of Kirkland Lake. He was a special education teacher, principal, and consultant with the Kirkland Lake Board of Education. He spent many years leading army cadets and was active in this volunteer field at the national level. Don was also very involved in the education and welfare of children with special needs. He worked hard to see their needs recognized and finally oversaw the founding of the first school for special needs children in Canada, first in a church basement and finally in a purpose-built school. He was honoured with membership in the Order of Canada in the early seventies. (Museum of Northern History)

SHUT DOWN. A large local landmark, the Teck Hughes Mine, shut down in 1968 after having produced precious metals since 1917. Its tally was 3,709,007 ounces of gold and 501,657 ounces of silver. Both headframes and most buildings were torn down over the next few years. One curiosity, a great, caved-in pit still remains in the location of the former mine, heavily fenced for safety.

STILL BUILDING. The Adams Mine is shown here, ready to begin production with giant open-pit excavations. The product sent south by rail to Hamilton would take the form of enriched-content iron pellets. The mine is closed now but some of the black pellets of iron may be still found. (Museum of Northern History)

MAKE SAFE. This scene was familiar around the gold town. Here, workers cap the shaft of what was once the second-largest mine in the camp. The Wright Hargeaves Mine milled 4,821,296 ounces of gold. The headframe was dismantled and the great shaft was covered to remove any hazard. Today, a liquor store sits on the site. (Museum of Northern History)

DEEP HOLES. Over its lifetime the Adams Mine carved huge pits in benches. Huge Euclid trucks drove along these benches to bring ore to surface. The pits were proposed as a possible dumping ground for Toronto's garbage. The one left could hold ten million tonnes; the right had capacity for seventeen million tonnes. (Museum of Northern History)

GONE NOW. The old Central School was still in place when the new Central School was completed (background). For a while, the old school housed the classrooms of Northern College before being demolished to make way for a new post office. (Museum of Northern History)

HILL SIDE. Central School was built on the site of the former Townsite Mine. The hill to the right was made up of broken rock and part of the building was over the 700-foot (now well capped) shaft. The author was principal of this school in the eighties. (Museum of Northern History)

NICE HOUSE. Harry Oakes supervised much of the construction of his Chateau. In the eighties, with funds from the Oakes family and the province, it was restored as a nursing home, and later opened as a museum with a gallery for exhibitions. (Museum of Northern History)

WELCOME FOLKS. As Kirkland Lake moved away from a solely mining image there was (and still is) an active push to sell the area for tourism. These unknown visitors admire signs meant to be informative and welcoming. (Museum of Northern History)

Eight
Sports, Music, and Culture

Kirkland Lake was established as a hard-working mining town. As the gold staking days faded away and people settled down to daily routines, so did they have time for amusements and hobbies. Sports played a large part in their lives and the mines were always ready to sponsor teams and special events. Many years later, famous athletes were hired by the most prominent mines and given jobs that would not interfere with their training and practice schedules. Every school had its own outdoor ice rink and skaters progressed to hockey. Today, the town is known as the cradle of hockey in Canada because so many local boys went on to play in the professional leagues.

The town became a true melting pot of cultures as men came from across the continent and Europe to work in the mines. In their spare time, many of them pursued their interest in music. There were always bands, some specializing in brass instruments and others less formal offering music of the day. Over time, there was always a variety of sports and musical entertainment available year-round.

TOUGH GUY. A bust of Sir Harry Oakes by an unknown sculptor sits on the mantelpiece in the main entrance room of the Chateau. It shows the strength of character of a strong man who found a mine, held on to it in the face of great difficulty, and made a fortune. (Museum of Northern History)

READY TO MARCH. The well-equipped Kirkland Lake Pipe Band stands in what are probably Central School grounds. The small mine in the background is the Townsite. (Museum of Northern History)

LUCKY COACH. This happy 1932 girls' softball team is posed with some small trophies garnered over the year. (Museum of Northern History)

BAND SCHOOL. This photograph of the Kirkland Lake Concert Band is undated. The members stand in front of the Kirkland Lake Collegiate and Vocational Institute where the front path awaits a layer of concrete. White shirts and miner's helmets make a nice contrast. (Museum of Northern History)

SOCCER TEAM. In this case the players were playing English football, which is commonly called soccer in Canada. The players were well supported by such

luminaries as Harry Oakes and Frank Duncan, likely with equipment costs. (Museum of Northern History)

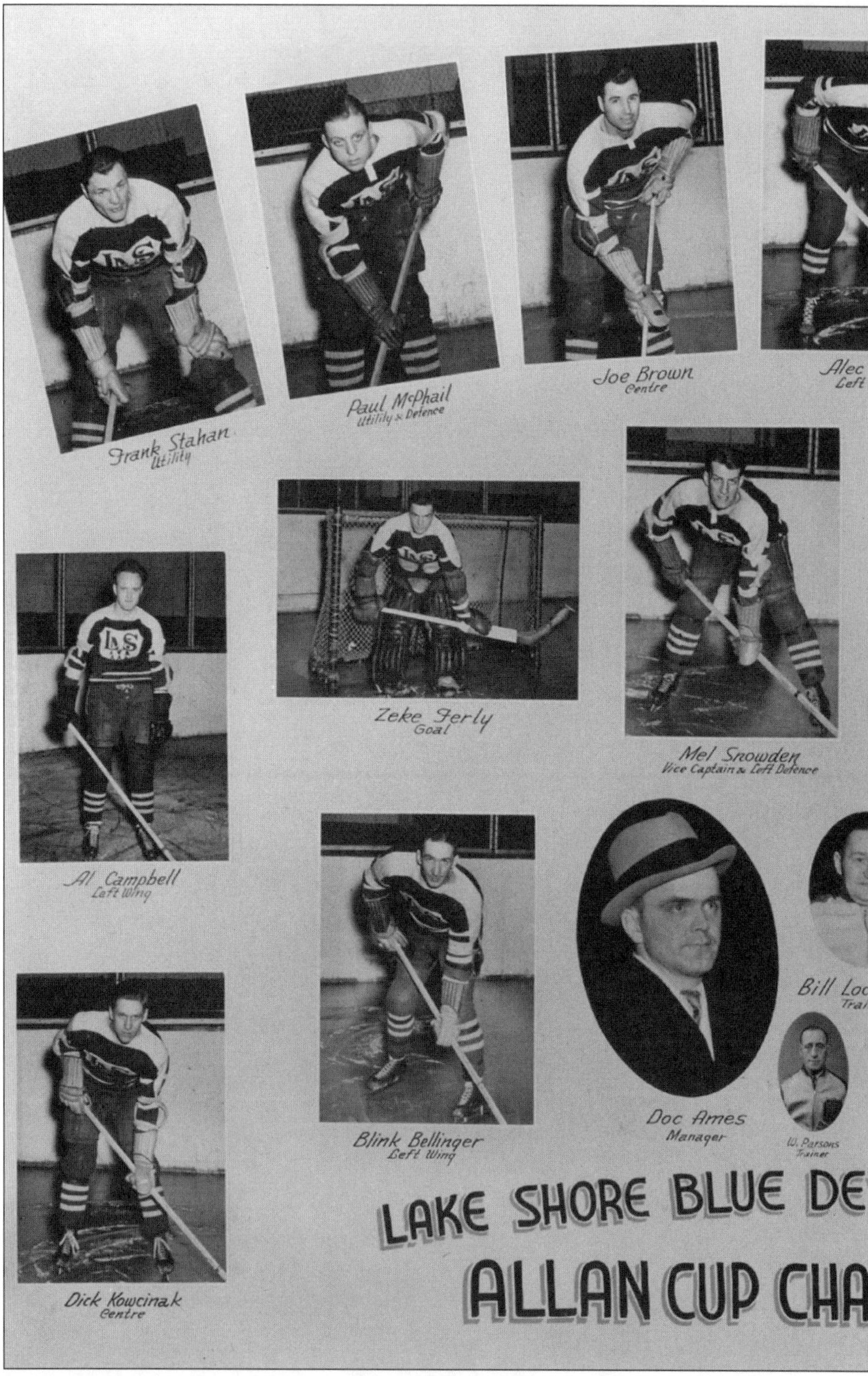

TOP SKATES. Among the players represented in this team picture are several who went on to play professional in the NHL. All were hardy men who had grown up

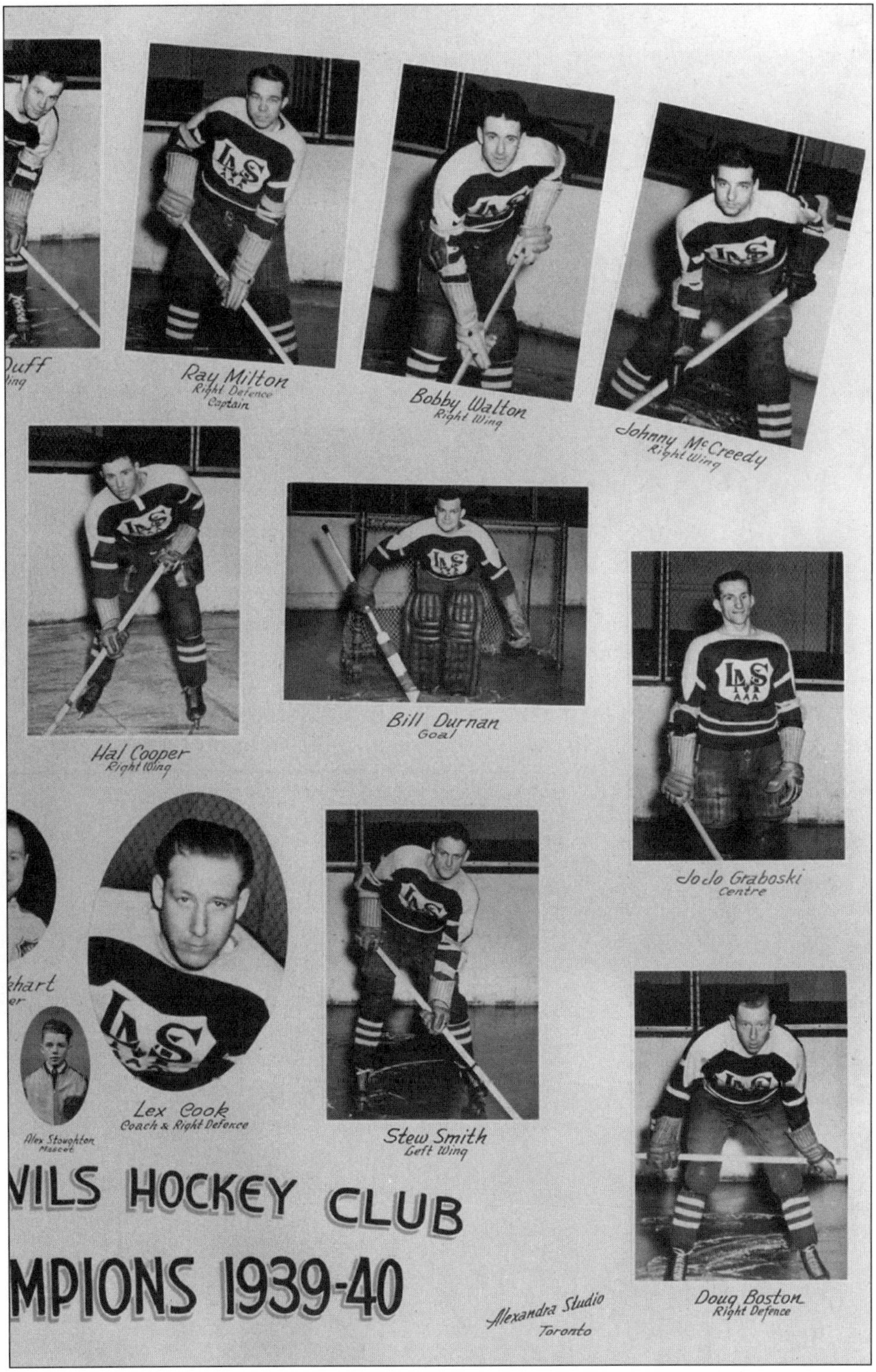

playing on outdoor rinks long before they played together as a team in the Kirkland Lake arena. (Museum of Northern History)

READY TO PLAY. Kirkland West School is gone now, but these players' images are preserved. Adults shown are: Ronald Goodman, Henry James, Walter Milbury, Principal J. Hutton, Norman Nankervis, and Edward Thomas. Boys shown are: Bobby Griffin (#9), David Powell (#13), Bobby Hattat (#14), Emile Novak (#6), Bobby Charlton (#1), Walter Milbury Jr. (#4), unknown (#7), Armand Perrier (#15), Beven Anderson (#5), Bobby Howes (#3), Norman Nankervis Jr. (#10), and Roy Thomas (#8).

SMART WORKERS. In 1946 members of the Junior Board of Trade decided to clean up the town. They paraded to the job with air cadets. Note many wore jackets and ties to sweep streets. (Museum of Northern History)

SPRING DANCE. The Septones played for the KLCVI 1950 Spring Dance. They were: Ted Roman (piano), Bob Allan (tenor sax), Terry Honer (alto sax), Jeff Scott (alto sax), Jim Rumball (drums), Carl Reinholt (trumpet), and Laurie Bower (trombone). (Museum of Northern History)

TOPS IN CANADA. In 1950 the Ramsay rink scooped the country and won the great Brier Trophy for curling. From left to right are, Bill Kenny, Tom Ramsay, Lynn Lynot Williamson, and Bill Weston. (Museum of Northern History)

WELL DRESSED. While hockey occupied the minds of most young boys, there was a great figure skating tradition in Kirkland Lake and many local ice stars went on to professional status. Perhaps the most well known is Toller Cranston of Swastika. The

annual ice show was a highlight in town and the 1945 event was no exception. (Museum of Northern History)

MUSIC MAKERS. The musical tradition in Kirkland Lake was well supported in the secondary school. Mr. Powell was a most successful music teacher and his bands

did well in competition. Many of the students seen here went on to success in music and professions. (Museum of Northern History)

BATTLER. In 1927 Harry Oakes became the patron of Niagara Falls area welterweight boxer Howard "The Kid" Williams. He put him up at the Chateau and got him a job. The Kid took the Canadian welterweight crown in 1930, won a silver medal at the British Empire Games, and represented Canada at the Los Angeles and Berlin Olympics. He repaid his backers by coming back to Kirkland Lake to train young boxers for a while. (Museum of Northern History)

POPPY DAY. The cenotaph was at one time located on Central School grounds. Here, Citizens Band conductor Sam Mason leads the musicians in the 1951 service. (Museum of Northern History)

GLAMOUR. In 1957 local ladies gave a fashion show rapt attention. Here, a self-assured unknown model struts her stuff in a fashionable long skirt and fur stole. (Museum of Northern History)

NIGHTCLUB ACT. The Park Lane Hotel had a cocktail bar downstairs and featured bands playing on weekends. In this 1954 shot of the Night Owls, the musicians did not hide their "pop" bottles. They were left to right, Dino Dottori (trumpet), Bill Zabarylo (accordion), Art Gabor (drums), Vic Zabarylo (alto sax), and Keith Conrad (piano). (Museum of Northern History)

FUN SKATE. Dick Duff was one of the great NHL players who came from Kirkland Lake, getting his game start at St. Jerome School and later St. Mike's. During his career from 1955 to 1972, the Stanley Cup winner with the Maple Leafs and Montreal Canadiens was also a coach with the Leafs and later a scout. He made many return trips to Kirkland Lake, both coaching and playing in charity games. Here, he rushes down the ice in memory of his former fast skating style. (Museum of Northern History)